YOUR KNOWLEDGE HAS VALUE

- We will publish your bachelor's and master's thesis, essays and papers

- Your own eBook and book - sold worldwide in all relevant shops

- Earn money with each sale

Upload your text at www.GRIN.com and publish for free

Bibliographic information published by the German National Library:

The German National Library lists this publication in the National Bibliography; detailed bibliographic data are available on the Internet at http://dnb.dnb.de .

Imprint:

Copyright © 2017 GRIN Verlag, Open Publishing GmbH
Print and binding: Books on Demand GmbH, Norderstedt Germany
ISBN: 9783668581906

This book at GRIN:

http://www.grin.com/en/e-book/380431/stresses-in-multiple-excavations-a-compa-rative-study

Arijit Ghosh

Stresses in Multiple Excavations. A Comparative Study

GRIN Publishing

Comparison of Analytical and Numerically computed load bearing behaviour of parallel drifts idealized both square and circular

Submitted in fulfilment of the requirements for Student Research Project to

The Institute of Mining

TU Clausthal

by

Arijit Ghosh

Master of Science (Mining Engineering)

Institute of Mining

Clausthal University of Technology

Abstract

Determining the stresses that exist in the rock mass opening has been a significant area of research in mining for a long time. The main reason for this is the concern of roof collapse of the excavation due to overlying pressure from the rock mass. This thesis is concerned with the state of stresses that exist in the rock mass in 2 conditions: 1. Stresses in rock before the mine openings; 2. Altered state of stresses after the excavations are made.

Theories have been formulated to calculate the pre-stressed state of the rock mass. But any sort of measurement invalidates the original condition of the intact rock. There are mainly 3 ways to study these different stress conditions in the rock mass.

> - Analytical solutions using pre-defined mathematical equations
> - Using numerical modelling to predict the stresses by duplicating the in-situ stress conditions.
> - In-situ measurements of the stress conditions in the rock mass.

One of the first assumptions we take while doing the measurements is assume that the rock mass is elastic and homogenous (single layer). There are also assumptions where the rock is considered viscous, plastic or a combination of these. But for this research project, I have considered the first assumption. This is done to ease up the subsequent calculations on the more complex characteristics of the rock mass.

In this research project, we attempt to do a comparison of the initial and final stressed conditions of the rock mass in a series of square and circular drifts between the analytical and the numerical solutions of the model. It must be noted that a completely accurate picture of the stress phenomena cannot be drawn because of the lack of knowledge of the physical properties of rock under field conditions.

The purpose of these project is to compare numerically calculated loading behaviour of rock mass to that of analytical solutions obtained from the same.

Table of Contents

List of Figures

1.0   Pre-stressed state of the rock mass

The pre-stressed state of the rock mass is influenced by the following factors:

- ➢ Weight of the overburden
- ➢ Depth of overburden
- ➢ Geological discontinuities (folds, faults etc.)
- ➢ Physical characteristics of the surrounding rock

A reasonable hypothesis for the initial stresses existing in underground rock before an excavation has been introduced was given by Mindlin in 1939. It was assumed that the stresses inside the earth at different depths maybe approximated by one of the 3 states of pressure. As shown in fig 1. (Caudle, 2007)

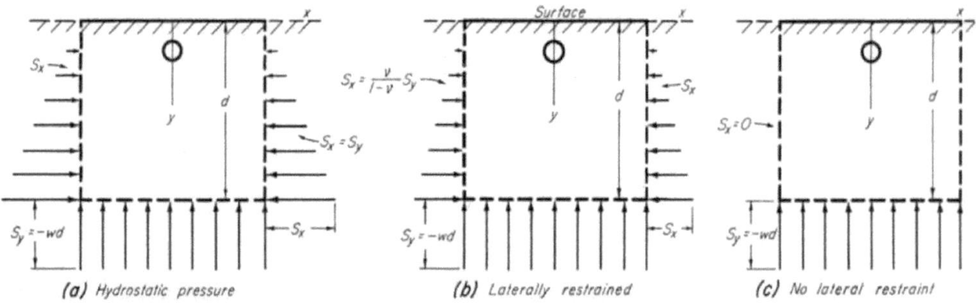

Fig. 1 : Assumed state of stress from any disturbing influence. (Caudle, 2007)

These cases represent the state of variation of the earth stresses. The actual initial stress condition lies between the 2 extreme values mentioned in the diagram. Hence, these 3 conditions mentioned in the Fig. are widely used in elastic analytical methods. Sometimes, however insignificant, soil erosion results in the decrease in the vertical stress on the rock mass. But the lateral pressure remains the same. Because of this, the resultant force pushes the rock mass together and can lead to mountain building.

Beyl (1952), obtained the state of stress in the rock mass at the surface and at depth by superposing three fields of pressure:

> Horizontal force due to orogenetic pressure of the rock mass.
> Vertical force due to the overlying weight of the rock mass.
> A hydrostatic pressure equal in all directions.

Since using these forces requires a knowledge of the orogenetic pressure of the rock mass, which we have not considered in our solution, we can use the theories of Mindlin to solve the problem in the analytical solution.

2.0   Stresses around excavations in solid homogenous materials

For the purposes of the project, we have selected the seam of coal and assumed it to be homogenous with no faults existing in the rock mass or the seam. The material is assumed to be elastic for the purposes of easier calculations.

The main aim of doing the stress analysis is to determine the existing stresses in the rock mass before the excavation, and the subsequent change and redistribution of the stresses after the excavations have been introduced.

Many early investigations conducted in obtaining the stress states of the rock mass was centred around the fact that a dome shaped space used to form around caved underground openings. When the opening is finished, the rock fails after a while and the original opening was converted to a dome shaped opening which re-established equilibrium in the rock mass. This simple observation gave rise to the dome theory in excavation engineering and is the cornerstone of our solution using the analytical method.

According to the dome theory, we assume that the rocks overlying an excavation are acted on by two forces only – cohesion and gravity. When the value of the gravitational force increases compared to the cohesion, there is roof failure forming an enlarging arch as shown in fig. 2 (Singh, 2005). Subsidence can occur if the dome reaches the surface.

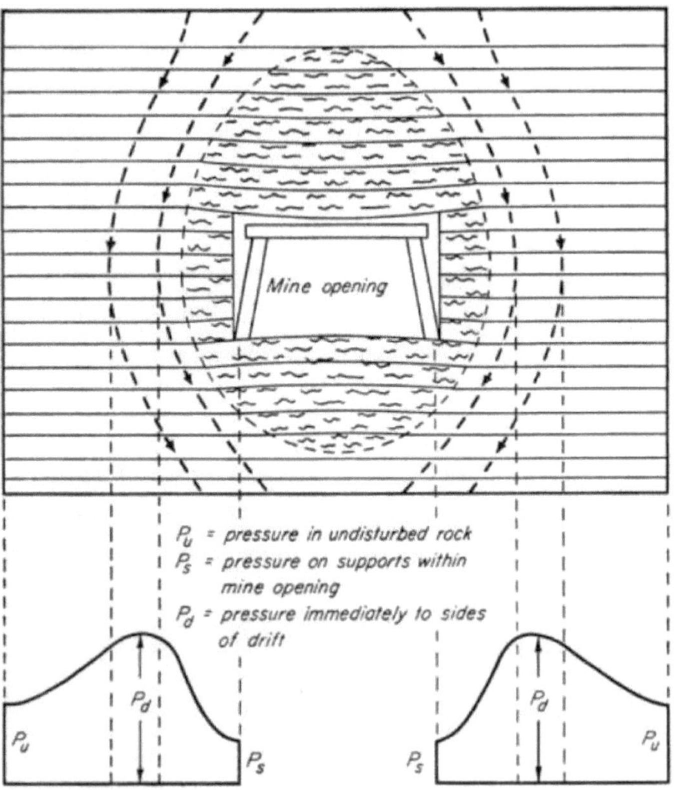

Fig. 2 : Pressure dome and stress trajectories around a drift (Caudle, 2007)

2.1 Stresses around single excavations

As the starting point of any calculation, we must first consider the effect of a single excavation on the underground and then move on to multiple excavations. This is of paramount importance as the pressure redistribution changes in different shapes as well as the number of excavations and the circular excavation seems like the most basic shape to start the analysis with.

The main objective of these problems is to achieve:

> Effect of different shapes in stress concentrations around the boundary of the excavation.
> The most stable shape to perform the excavation to avoid failure of the rock
> Determine in situ stress around the mine openings

2.1.1 Circular Excavation

The 3 states of pressure that existed before the excavation was made is used to solve the problem of the stress due to gravity in the circular excavation. During these calculations, we consider the length of the tunnel to be infinite compared to the cross-sectional diameter of the excavation. This is also one of the preliminary conditions mentioned in the problem.

Panek (1952), postulated that the maximum tensile and compressive stresses occur in the boundary of the opening. The roof and floor of the excavation are in tension and the ribs are in compression. But when the lateral pressure is greater than one half the vertical pressure, the tangential stress becomes compressive. Therefore, the Poisson's ratio is important because it determines the lateral pressure (Caudle, 2007).

2.1.2 Square excavation

In square excavations, the stress concentrations are a little varied. It is a more unstable configuration, and hence, not widely used in mines. The maximum stress concentrations are around the edges of the mine pillars, on the roof. The floor sometimes is subjected to tensile stresses rather than compressive stress. This leads to uneven stress distribution and roof collapse. While making square pillars, we must keep in mind the cross-sectional area of the roof so that it can withstand the overlying rock mass. Most of the load measurements are done in-situ in case of a square excavation (less accuracy through analytical solutions).

3.0   Analytical solutions around multiple circular and square excavations

While doing analytical calculations we assume certain characteristics of the rock mass. The first one being that the rock is elastic and isotropic. The mineral group where the excavation is taking place is assumed to be a coal seam which is at a depth of 100 m. The initial conditions for finding out the solutions of stresses are mentioned in Fig. 3.

Horizontal *Principal* Stress (σh)	2.38 MPa
Vertical *Principal* Stress (σn)	2.38 MPa
Ratio between horizontal & Vertical **Primary** Stresses (K0)	1
Radius of excavation (a)	5 m
Support Resistance (pi)	0 MPa
Angle of rotation against x-axis (φ)	0 degree
Depth (d)	100 m
Density (p)	2.5 t/m3
Vertical Stress (σv)	2.38 MPa
Distance from excavation (r)	
Poisson Ratio (v)	0.3
Young's Modulus (E)	5000 MPa

Fig. 3 : Initial conditions assumed for the analytical solution

As we can see from fig. 3, the support resistance is assumed to be zero. When we have not made the excavations, the initial vertical pressure of the entire rock mass along the middle of the coal seam is calculated as 2.38 MPa.

The equations used to obtain the analytical solutions are displayed in Fig. 4. The first two equations are for the radial and tangential stresses respectively. The last two equations are for the radial and tangential strains respectively. The symbols mentioned in the equations are all described in fig. 3 with the initial conditions.

$$\sigma_r = \frac{a^2}{r^2} * p_i + \frac{\sigma_v}{2}\left[2 * \left(1 - \frac{a^2}{r^2}\right)\right]$$

$$\sigma_\varphi = -\frac{a^2}{r^2} * p_i + \frac{\sigma_v}{2}\left[2 * \left(1 + \frac{a^2}{r^2}\right)\right]$$

$$\varepsilon_r = -\frac{\sigma_v * a^2}{4G * r^2} * 2 + \frac{(1 + v) * a^2}{E * r^2} * p_i$$

$$\varepsilon_\varphi = \frac{\sigma_v * a^2}{4G * r^2} * 2 - \frac{(1 + v) * a^2}{E * r^2} * p_i$$

Fig. 4 : Equations for stress and strain in circular excavations (Düsterloh,2016)

The equations have been simplified due to the value of K_0 being 1 and some of the terms cancelling out.

The results of the analytical solution are in table 1. Note that this solution is for circular excavations only. The solution for square excavations is also assumed to be same as the square excavations are idealised as circular and the dimensions remain the same. The values we get from these equations are a result of 2 excavations on either side of the pillar. the values mentioned here are a resultant of the stresses due to both the excavations on the pillar. To find the resultant stresses, we simply add the stresses from both the excavations and then subtract the vertical stress from the result to obtain the stress on each part. The analytical solutions are only an assumption and not related to the actual field conditions.

The graphs of the stresses and strains will be analysed in the later part of the study.

Radial dist. (exc. 1) (m)	Radial Stress (MPa)	Tangential stress (MPa)	Radial Strain	Tangential Strain	Equivalent Stress (MPa)	Equivalent Strain
5.25	-0.052	4.812	-0.00063	0.00063	4.2127	0.00147
5.75	0.287	4.473	-0.00054	0.00054	3.6246	0.00129
6.25	0.542	4.218	-0.00048	0.00048	3.1834	0.00115
6.75	0.735	4.025	-0.00043	0.00043	2.8489	0.00105
7.25	0.882	3.878	-0.00039	0.00039	2.5946	0.00097
7.75	0.993	3.767	-0.00036	0.00036	2.4026	0.00092
8.25	1.075	3.685	-0.00034	0.00034	2.2606	0.00088
8.75	1.133	3.627	-0.00032	0.00032	2.1603	0.00085
9.25	1.170	3.590	-0.00031	0.00031	2.0963	0.00084
9.75	1.188	3.572	-0.00031	0.00031	2.0650	0.00083
10.25	1.188	3.572	-0.00031	0.00031	2.0650	0.00083
10.75	1.170	3.590	-0.00031	0.00031	2.0963	0.00084
11.25	1.133	3.627	-0.00032	0.00032	2.1603	0.00085
11.75	1.075	3.685	-0.00034	0.00034	2.2606	0.00088
12.25	0.993	3.767	-0.00036	0.00036	2.4026	0.00092
12.75	0.882	3.878	-0.00039	0.00039	2.5946	0.00097
13.25	0.735	4.025	-0.00043	0.00043	2.8489	0.00105
13.75	0.542	4.218	-0.00048	0.00048	3.1834	0.00115
14.25	0.287	4.473	-0.00054	0.00054	3.6246	0.00129
14.75	-0.052	4.812	-0.00063	0.00063	4.2127	0.00147

Table 1: Analytical solutions of the circular excavations and idealised square excavations

While using numerical modelling software, we must keep in mind first the boundary conditions of the model. For this project, we have used FLAC 3D to analyse the loading behaviour of the rock mass. It is a rock mechanics software used to model these excavations and simulate the approximate conditions as one might find in the real excavation. To write the code for the boundary conditions as well as to model this, we have used the FISH language which is an indigenous programming language exclusive to FLAC 3D.

During the start of this project, the first thing to do is to learn the library of FISH functions which help in creating the boundary conditions and other functions. Without getting in detail into the language and the programming part, we will concentrate on the execution part of the numerical modelling. After we create the model including the rock mass and the mineral, we provide different characteristics to each of them including the young's modulus, density and strength of the rock mass or the mineral. Based on these properties, we get an initial model of the excavation as shown in Fig. 5 where the groups are mentioned.

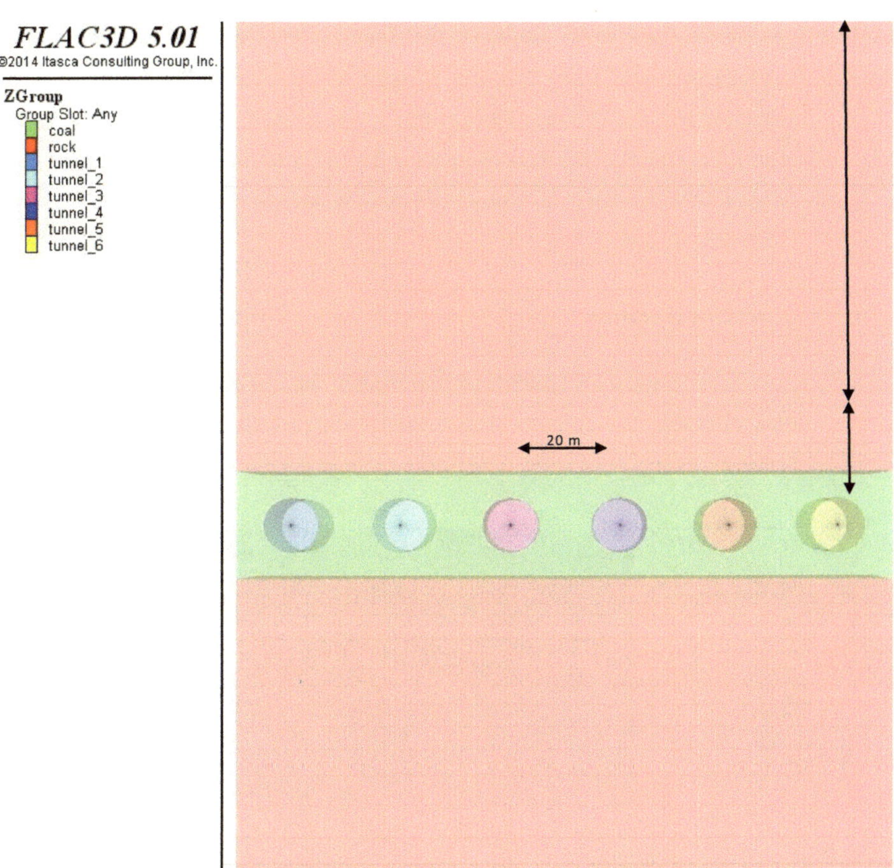

Fig. 5 : Initial zone assignment for simulation

Once these are discretised, we move on to applying the load on the model under the given boundary conditions. The model is discretised into individual zones on which the loads and other properties are applied. For our project, we have taken a total of 12000 zones. Of these, 4800 zones are assigned to the rock and coal group each, and the remaining 2400 zones equally divided among the 6 tunnels. In this project, we have assumed a mechanical elastic model. First the load is applied to the unexcavated rock mass. After that, the excavations are made and the load is applied on the excavated rock mass for both the shapes. The results are summarised usually in colour scale to get a glimpse of the load distribution. The list function in FLAC 3D is used to get the values of loading on specific points in the model so that we can compare it to the analytical solutions obtained above. The results of numerical modelling on both the shapes are mentioned below.

4.1   Circular Excavation

The results were analysed from the model and are represented in tabular form in table 2. A glimpse of how the results look like is shown in Fig. 6. It represents the equivalent stress of the model in colour scale.

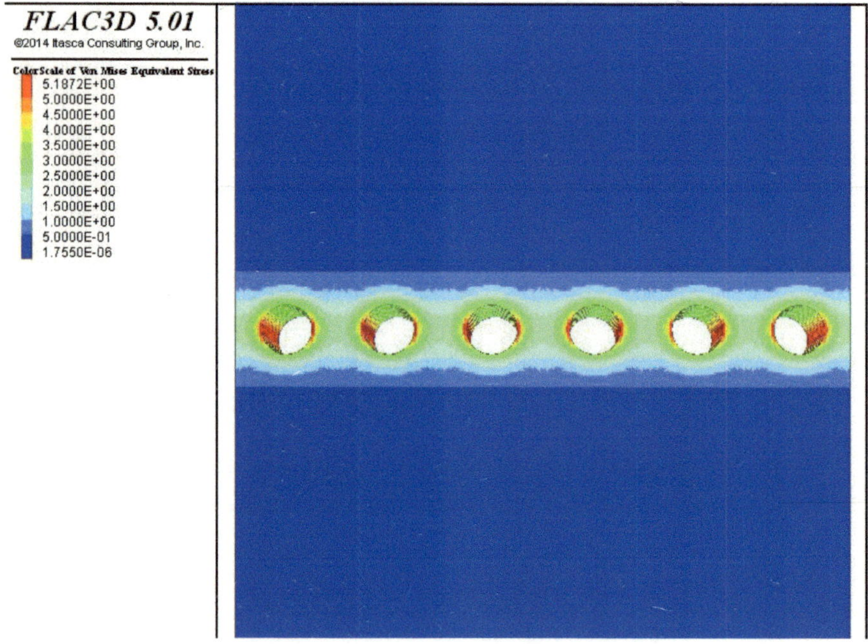

Fig. 6 : Equivalent stress of the FLAC 3D model for circular excavations

The points that we use for the readings of the numerical simulation are all located at the centre of the respective elements of the model. So, the stress we measure is at the centroid of each element in the pillar because it is an average Stress applied to the whole element in the pillar.

Radial dist. (exc. 1) (m)	Radial Stress (MPa)	Tangential stress (MPa)	Radial Strain	Tangential Strain	Equivalent Stress (MPa)	Equivalent Strain
5.25	0.299	6.213	-0.00437	0.00252	5.184	0.00201
5.75	0.725	5.527	-0.00359	0.00201	4.213	0.00202
6.25	1.004	5.062	-0.00305	0.00167	3.568	0.00164
6.75	1.193	4.735	-0.00268	0.00144	3.116	0.00139
7.25	1.326	4.501	-0.00241	0.00128	2.796	0.00108
7.75	1.419	4.333	-0.00222	0.00117	2.564	0.00099
8.25	1.485	4.213	-0.00209	0.00109	2.401	0.00093
8.75	1.529	4.13	-0.00199	0.00103	2.292	0.00089
9.25	1.557	4.078	-0.00193	0.001	2.221	0.00086
9.75	1.57	4.053	-0.00191	0.000987	2.186	0.00085
10.25	1.57	4.053	-0.00191	0.000987	2.22	0.00085
10.75	1.557	4.078	-0.00193	0.001	2.289	0.00086
11.25	1.53	4.13	-0.00199	0.00103	2.401	0.00089
11.75	1.485	4.213	-0.00209	0.00109	2.564	0.00093
12.25	1.419	4.333	-0.00223	0.00117	2.793	0.00099
12.75	1.326	4.502	-0.00242	0.00128	3.113	0.00109
13.25	1.194	4.736	-0.00268	0.00144	3.564	0.00121
13.75	1.004	5.062	-0.00306	0.00168	4.213	0.00139
14.25	0.726	5.528	-0.00359	0.00201	5.187	0.00164
14.75	0.299	6.214	-0.00438	0.00252	5.188	0.00202

Table 2: Solutions from FLAC 3D for multiple circular excavations

4.2   Square Excavation

For the square excavation, the results for the equivalent stress look like as shown in Fig. 7. The rest of the stresses and strains are listed in table 3.

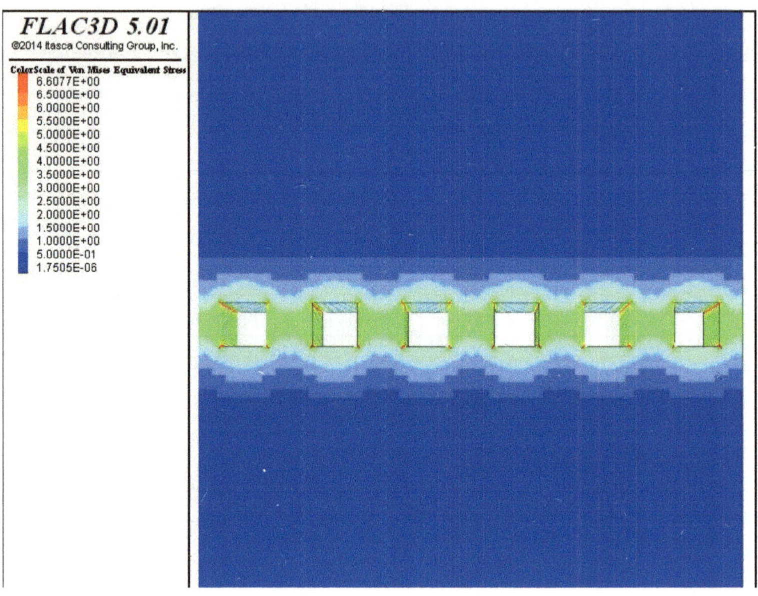

Fig. 7 : Equivalent stress of FLAC 3D model for square excavations

Radial dist.	Radial Stress	Tangential stress	Radial Strain	Tangential Strain	Equivalent Stress	Equivalent Strain
5.25	0.299	6.214	-0.00438	0.00252	3.631	0.00141
5.75	0.726	5.528	-0.00359	0.00201	3.813	0.00148
6.25	1.004	5.062	-0.00306	0.00167	3.931	0.00153
6.75	1.194	4.736	-0.00269	0.00144	3.982	0.00155
7.25	1.326	4.502	-0.00242	0.00129	3.937	0.00154
7.75	1.419	4.334	-0.00223	0.00117	3.944	0.00153
8.25	1.485	4.213	-0.00209	0.00109	3.937	0.00151
8.75	1.529	4.131	-0.00199	0.00104	3.83	0.00149

9.25	1.557	4.079	-0.00194	0.001	3.837	0.00147
9.75	1.571	4.053	-0.00191	0.00099	3.793	0.00146
10.25	1.571	4.053	-0.00191	0.00098	3.759	0.00146
10.75	1.557	4.079	-0.00194	0.001	3.793	0.00147
11.25	1.529	4.131	-0.00199	0.00104	3.83	0.00149
11.75	1.485	4.213	-0.00209	0.00109	3.884	0.00151
12.25	1.419	4.334	-0.00223	0.00117	3.944	0.00153
12.75	1.326	4.502	-0.00242	0.00129	3.98	0.00155
13.25	1.194	4.736	-0.00269	0.00145	3.982	0.00153
13.75	1.004	5.026	-0.00306	0.00168	3.924	0.00152
14.25	0.726	5.528	-0.00359	0.00201	3.809	0.00149
14.75	0.299	6.214	-0.00438	0.00252	3.631	0.00141

Table 3: Solutions from FLAC 3D for multiple square excavations

A detailed view of the radial and tangential stresses and strains for the square excavation are outlined in **Appendix A**. Subsequently, the stresses and strains for the circular excavation are outlined in Appendix B.

5.1   Analytical solution

When analysing the results of the analytical calculations, we get a graph as shown in fig. 8.

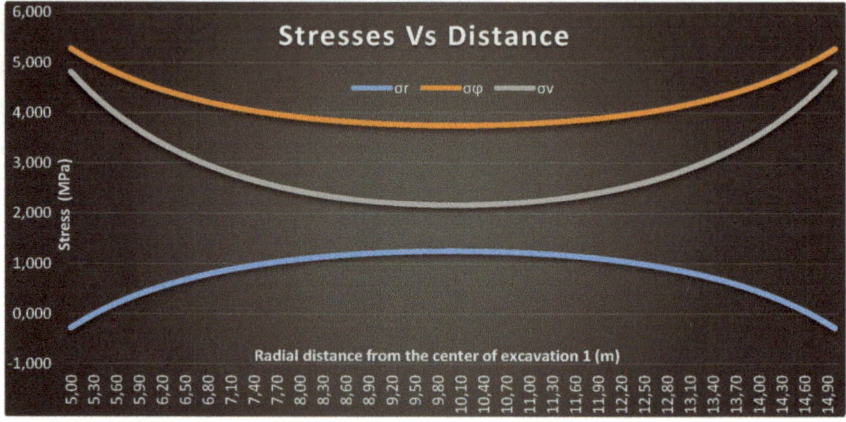

Fig. 8 : Stress curves for a Circular and idealised square excavation (analytical)

The radial distances are in steps of 5cm as opposed to the table where they are in steps of 50cm. This is because we used only the relevant readings in the middle to compare it with the numerically calculated stresses from FLAC 3D. For the comparisons, we start our values from 5.25m because the readings from FLAC3D are at the centroid of their respective elements. As expected in the analytical solution, the curve we obtain is quite isotropic and smooth because it does not represent the actual field conditions. The radial stresses (σ_r) are quite low as compared to the tangential stresses (σ_φ). The resultant stress is therefore somewhere in the middle (Grey curve) because of the vertical stress from the rock mass above.

For the strain graph, we can say that the graph looks smooth and isotropic just like the stress graph. This is shown in Fig. 9.

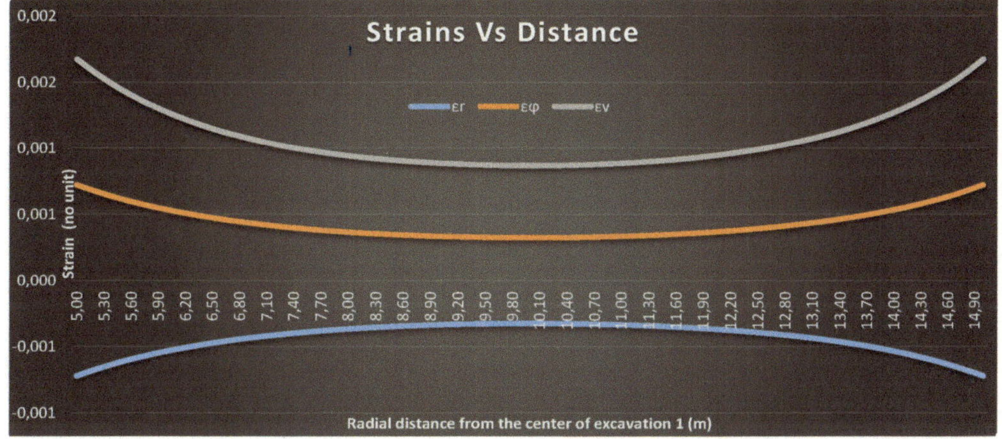

Fig. 9 : Strain Curves for Circular and idealised square excavations (analytical)

5.2   Numerical solutions for circular and square excavation

The graph of the stresses for the circular excavation is shown in Fig. 10. We compare it to the stresses obtained from the square excavation which are in Fig. 11. The first thing we observe here is the relative uniform distribution of the equivalent stress in case of circular excavation compared to the square excavation (grey curve). But both curves are not isotropic as we get from the analytical solution. There is a high stress value tending to 4 MPa for the square pillar as compared to the curved pillar where it is tending to 2.5 MPa.

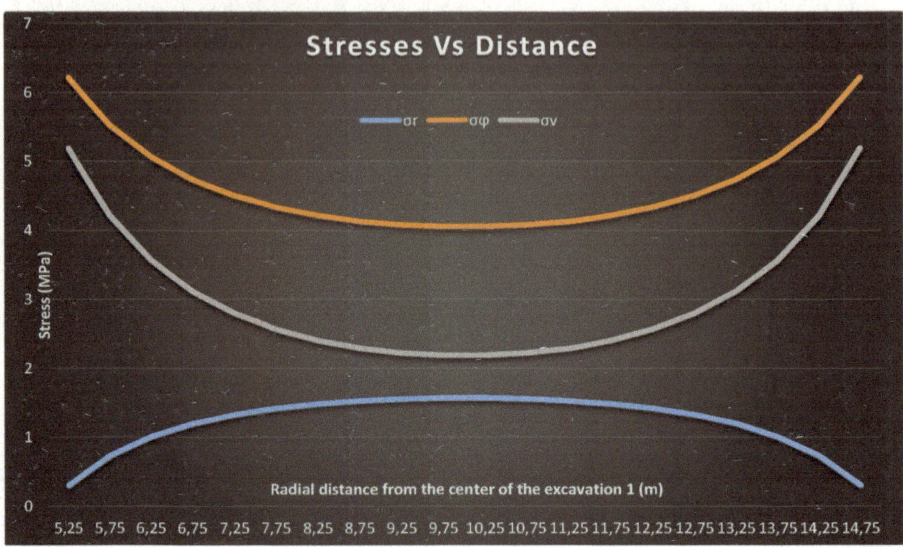

Fig. 10 :   Stress curves for numerically calculated data for circular excavation

This Low stress value leads to more stability in the pillar to withstand rock mass loads. The strain curves for the circular and square excavations are shown in Fig. 12 and Fig. 13 respectively.

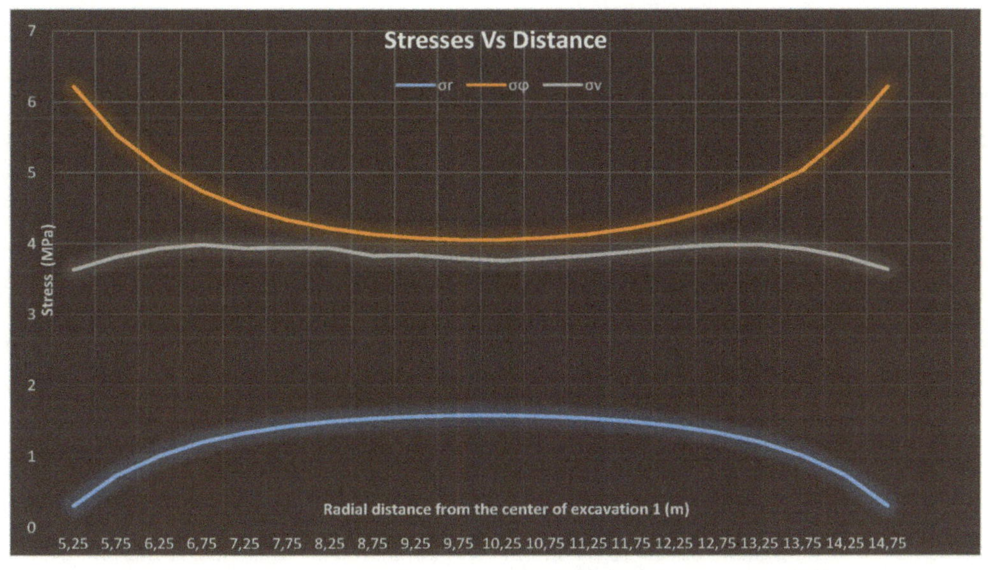

Fig. 11 :   Stress curves for numerically calculated data for square excavation

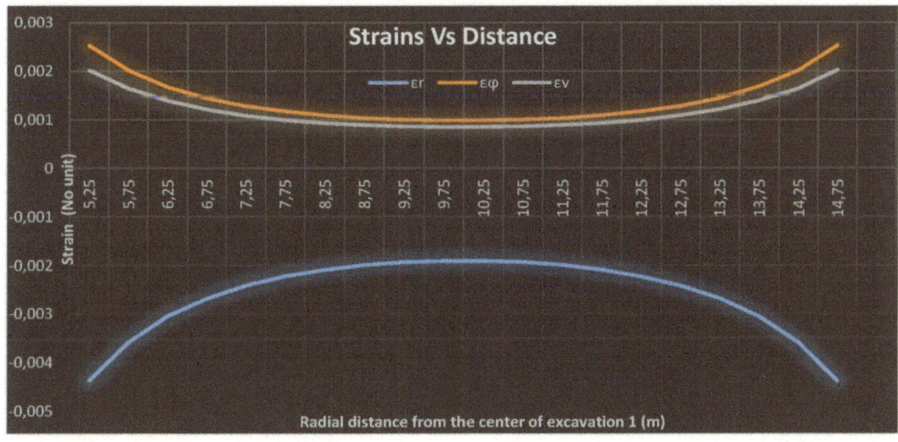

Fig. 12 :   Strain curves for numerically calculated data for circular excavation

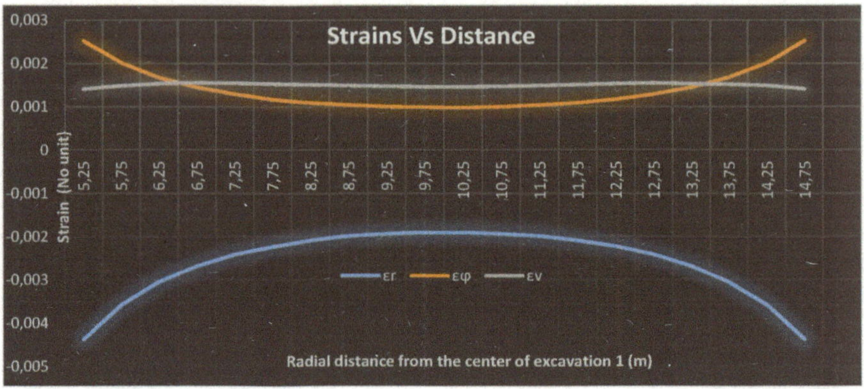

Fig. 13 :   Strain curves for numerically calculated data for square excavation

From both these curves, we can see that there is an overlap between the equivalent strain (ε_r) and the tangential strain (ε_φ). This is due to the cancelling out of the tangential strain with negative radial strain meaning there is radial displacement in the opposite direction. This is a good thing as it reduces the overall displacement of the rock mass.

5.3   Comparison of equivalent stresses and strains

After analysing the stress and strain curves of both these excavations in different models, we compare their equivalent or resultant stresses and strains to see how stable or unstable each model is and how accurate the analytical solutions are as compared to the numerical solutions. The graph we get for equivalent stresses is shown in Fig. 14. Subsequently, the graph for equivalent strains in shown in Fig. 16.

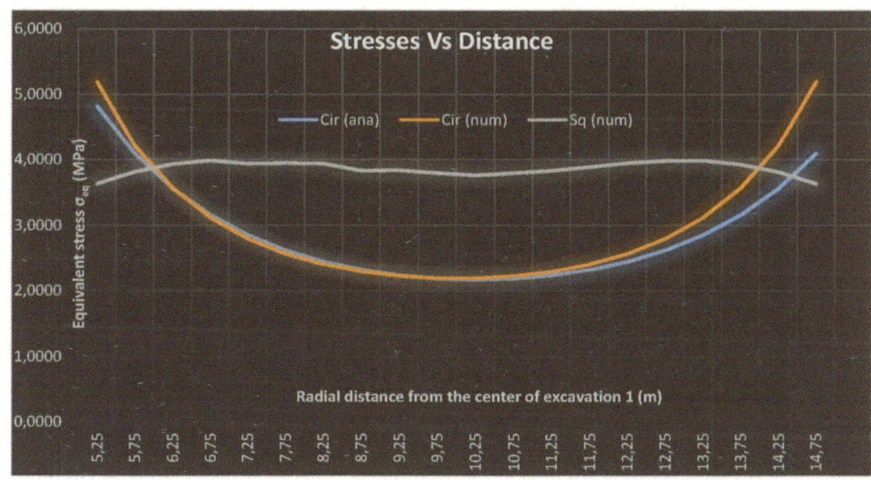

Fig. 14 :   Comparison of equivalent stresses in numerical and analytical solutions

In case of Fig. 14, we observe that the numerically calculated stress for circular excavation is almost like the analytically calculated stress value. But the values for the square excavation is quite different with the values being on the higher side tending to 4 MPa almost throughout the whole pillar. When we do a regression analysis for the numerically calculated values we get the following results as shown in Fig. 15. For the purposes of this project, the regression analysis has been done using the analyse function in Microsoft excel. It can compare any 2 sets of given values for their degree of correlation.

Circular Excavation	
Regression Statistics	
Multiple R	0.925347181
R Square	0.856267405
Adjusted R Square	0.848702532
Standard Error	0.43420508
Observations	21

Square Excavation	
Regression Statistics	
Multiple R	0.538534487
R Square	0.290019394
Adjusted R Square	0.252651993
Standard Error	0.098710682
Observations	21

Fig. 15 : **Regression analysis (stress) of numerical values with analytical values**

After regression analysis, the value of interest to us is the adjusted R square values. It gives a numerical value of the degree of correlation between 2 sets of values. The adjusted R square value in case of circular excavation is around 0.85, which is an acceptable value for the correlation. This means that the analytical solutions are close to the simulation values (although 0.9 to 1 is the standard accepted norm). On the other hand, the adjusted R square value for the square excavation is around 0.25 which shows that idealising the square pillars as circular pillars as square pillars gives a value which is not even close to the simulated value, which might be the case in field conditions. Therefore, the square excavation values must not be taken from these analytical solutions as they are highly inaccurate.

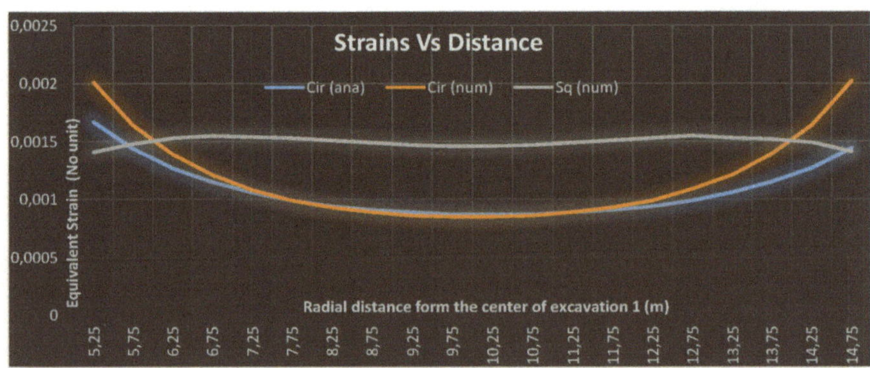

Fig. 16 : **Comparison of equivalent strains in numerical and analytical solutions**

Now we come to the regression analysis of the strain measurements in both the cases. The results are shown in Fig. 17.

Circular Excavation			Square excavation		
	Regression Statistics			*Regression Statistics*	
Multiple R	0.973274992		Multiple R	0.529551361	
R Square	0.947264209		R Square	0.280424644	
Adjusted R Square	0.944488641		Adjusted R Square	0.242552257	
Standard Error	6.09955E-05		Standard Error	0.000225311	
Observations	21		Observations	21	

Fig. 17 :   Regression analysis (strain) of numerical values with analytical values

The regression analysis of strains for circular excavations gives an adjusted R square value of 0.94 which is a very accurate value for the strain. Hence, the analytical solutions gives us an accurate picture of the strains in field conditions. On the other hand, the Value for the square excavation is 0.24 which is again an indicator that it is not an accurate representation of the field conditions.

6.0   Conclusion

In conclusion, the stresses that we calculated are for mainly isotropic and elastic conditions of rock masses. The effect of introduction of mine openings for simple geological conditions have been solved due to the ease of modelling these conditions. The analytical solutions for the circular excavation is quite accurate in simulating the field conditions. Such is not the case for square excavations.

It is observed from the numerical models simulated by FLAC 3D that the stress on the pillars increase slightly with the addition of more excavations. This happens until the 6th excavation is introduced. There is hardly any difference in the stress distributions on the 6th and 7th pillar. So additional excavations do not effect the stress concentrations of the other pillars.

We have not considered the effect of geological features which might exist in the rock mass such as folds, faults, fissures have not been considered in this project. Introducing these will change the conditions under which we evaluate the rock mass, such as introducing supports to combat the instability of the overlying rock mass.

Comparing circular and square excavations, we see that the circular excavation is much more stable since the stress distribution is fairly isotropic and the centre of the pillar has the least stress compared to the edges where it increases slowly. In case of the square pillar, high stresses in the middle of the pillar as well as on the edges lead to a highly unstable rock condition which may lead to collapse of the immediate roof at any time. So, it is advisable to use an excavation shape largely resembling circular excavation but also having a stable floor for the vehicles to travel in.

7.0   References

Beyl, Z.S., (1951). "the prestressed state of the earth's crust, its causes and preservation,"

Caudle, R.D. and Clark G.B., (2007). "Stresses around Mine openings in some simple Geologic structures".

Düsterloh, U., (2016). "Advanced rock mechanics lectures" pp. 333.

Mindlin, Raymond D., (1939). "stress distribution around a tunnel" pp. 619-642.

Panek, Louis A., (1951), "stresses about mine openings in a homogenous rock body".

Singh, R.D., (2005). "Principles and practices of modern coal mining" pp. 449.

8.1   Appendix A: stress and strain distribution around a square excavation

The radial and tangential stresses and strains whose resultant equivalent stress distribution we observed in fig. 6 are detailed here from FLAC 3D. The distribution is shown by zooming in on one single excavation (one section) to get a detailed view of the simulated stress distribution around the tunnel.

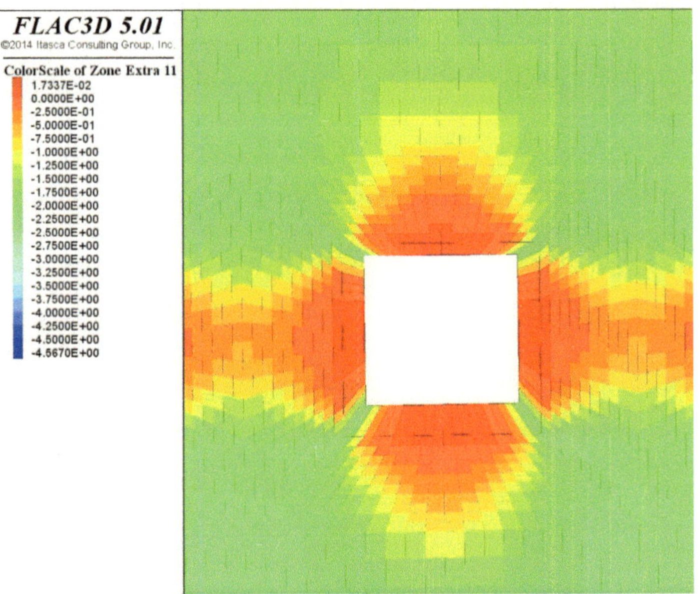

Radial stress around single square excavation

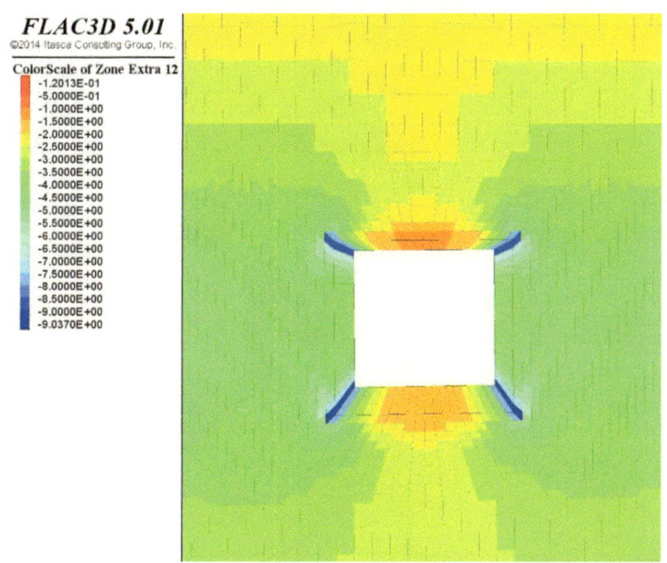

Tangential stress around single square excavation

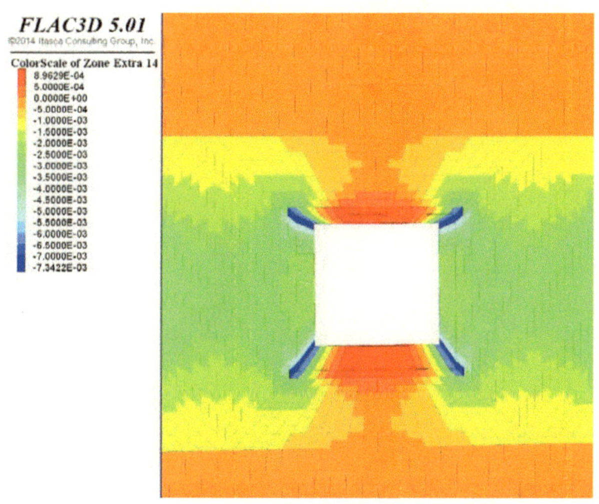

Radial strain around single square excavation

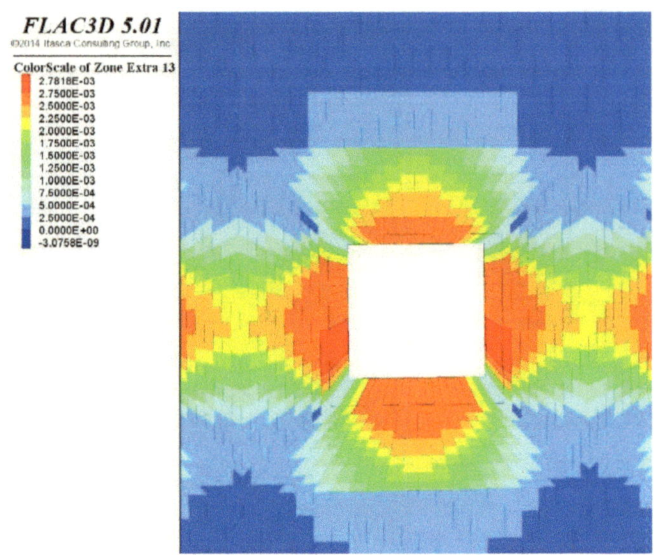

Tangential strain around single square excavation

8.2 Appendix B: stress and strain distribution around a circular excavation

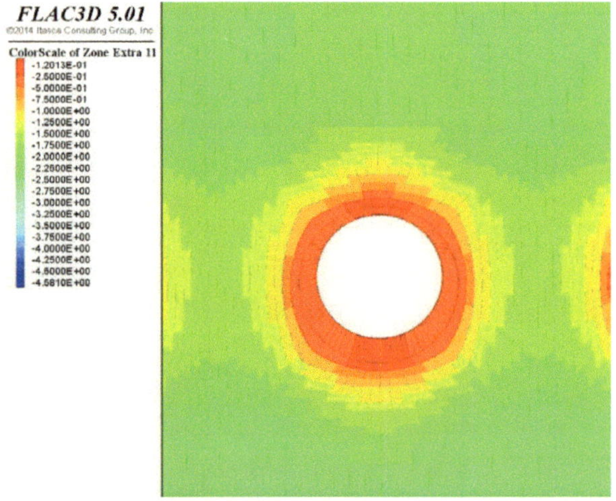

Radial stress around single circular excavation

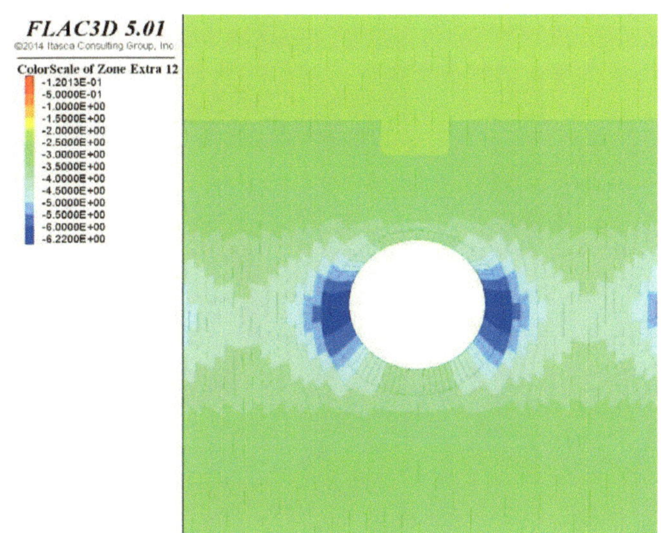

Tangential stress around single circular excavation

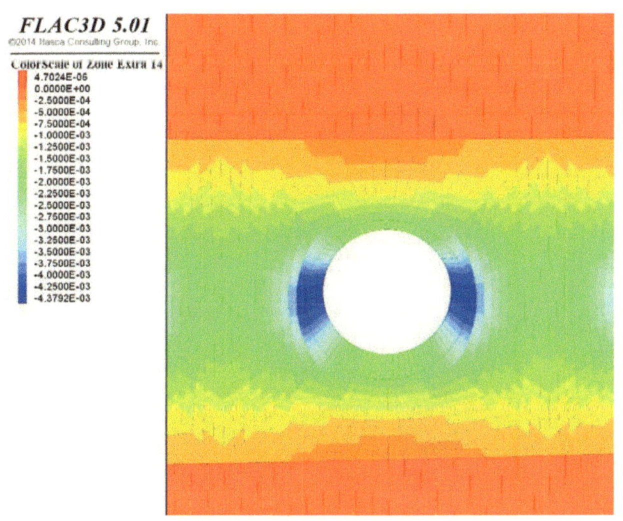

Radial strain around single circular excavation

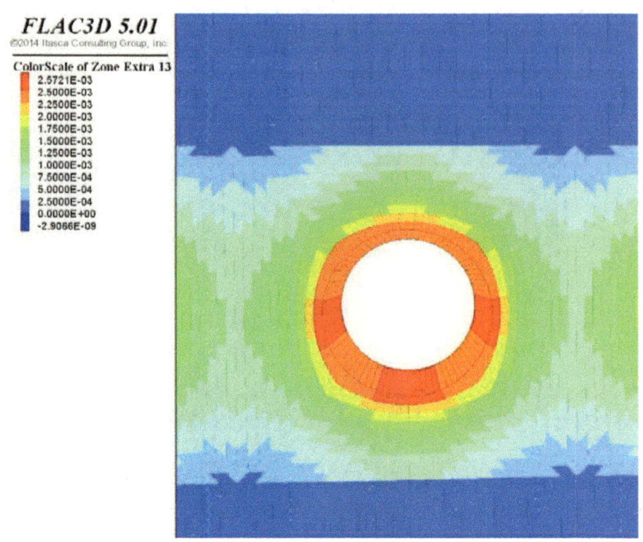

Tangential strain around single circular excavation